FARM MACHINERY

R. J. Stephen

Franklin Watts

London   New York   Sydney   Toronto

© 1986 Franklin Watts Ltd

First published in Great Britain
   1986 by
Franklin Watts Ltd
12a Golden Square
London W1R 4BA

First published in the USA by
Franklin Watts Inc
387 Park Avenue South
New York
N.Y. 10016

First published in Australia by
Franklin Watts
14 Mars Road
Lane Cove
2066 NSW

UK ISBN: 0 86313 418 1
US ISBN: 0-531-10186-X
Library of Congress Catalog Card
Number 85-52092

Printed in Italy
by Tipolitografia G. Canale & C. S.p.A. - Turin

Designed by
Barrett & Willard

Photographs by
Alfa-Laval
Australian News & Information Bureau
N. S. Barrett Collection
Bomford & Evershed
Case International
County Tractors
Fiatagri
FMC
Four Seasons Publicity
Howard Farmhand
Massey Ferguson
SKH
Smallford Planters
Sperry New Holland
Wilder Engineering
Wright Rain
ZEFA

Illustration by
Rhoda & Robert Burns

Technical Consultant
Steven Vale

Series Editor
N. S. Barrett

Contents

Introduction

Machinery is used for all kinds of farm work. There are machines for preparing the soil and machines for helping the crops to grow. Other machines harvest the crops.

On modern farms, machines are also used in the farming of livestock, such as cattle or sheep. Cattle, for example, may be milked and fed with the help of special machinery.

△ The tractor is the most used farming machine. It pulls all kinds of other farm machines and implements.

In developed countries, such as those of Europe or North America, farmers use machines for most agricultural work. They employ few farm laborers.

But in third world countries, in Asia and Africa for example, workers still do most of the jobs on the land. Gradually, machinery is being introduced in these countries to increase production.

△ This forage harvester collects and chops up cut grass and blows it into the trailer. The grass is stored as silage to make cattle feed.

The combine harvester

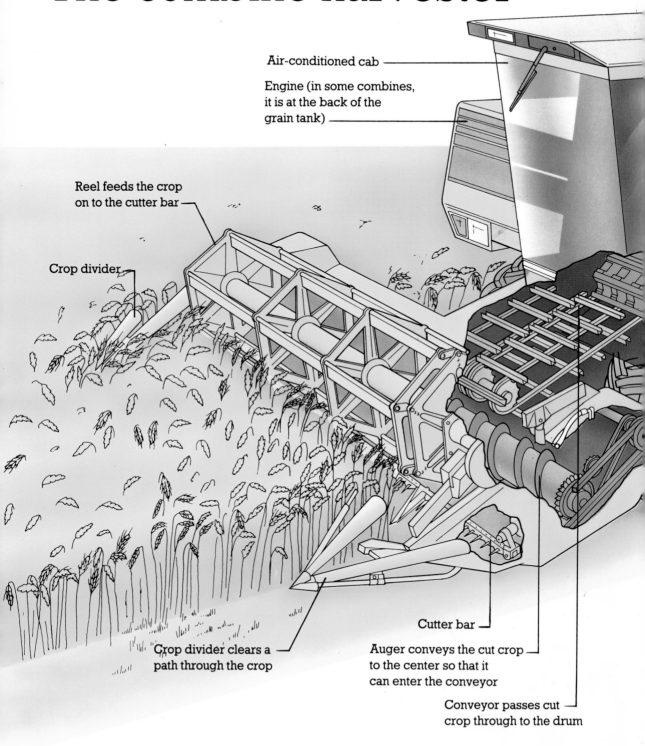

Air-conditioned cab

Engine (in some combines, it is at the back of the grain tank)

Reel feeds the crop on to the cutter bar

Crop divider

Crop divider clears a path through the crop

Cutter bar

Auger conveys the cut crop to the center so that it can enter the conveyor

Conveyor passes cut crop through to the drum

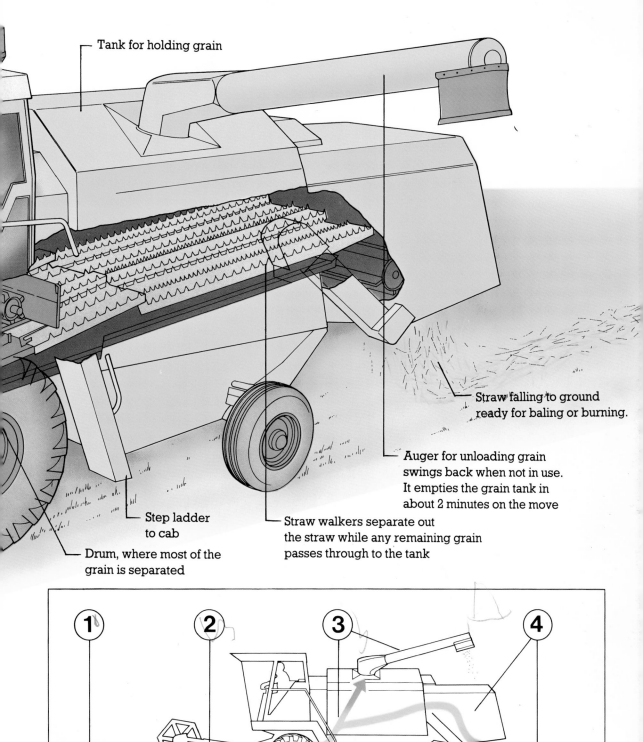

Tank for holding grain

Straw falling to ground
ready for baling or burning.

Auger for unloading grain
swings back when not in use.
It empties the grain tank in
about 2 minutes on the move

Straw walkers separate out
the straw while any remaining grain
passes through to the tank

Step ladder
to cab

Drum, where most of the
grain is separated

①
②
③
④

1 Crop being harvested.
2 Stalks cut and conveyed to drum.

3 Grain passed through to tank.
4 Straw passed out at back.

Tractors

The tractor is the workhorse of the modern farm. It has large wheels and a powerful engine. It does most of the pushing and pulling work in the fields, as well as the lifting and carrying. Many tractors have four-wheel drive to give extra grip on hills and in wet conditions.

Other machines, or implements such as plows or balers, are attached to tractors.

△ A powerful tractor pulling a plow.

▷ A crawler (top left) is a tractor with steel tracks instead of wheels. Crawlers are used when the ground is rocky and would tear rubber tires. Large tires help a tractor (top right) to move on soft or loose ground. A tractor with a loading attachment for silage or manure (bottom).

Preparing the soil

Soil has to be prepared for growing crops. Plows and cultivators are used to dig up and mix the earth. This kills the weeds and helps the air and water in the soil circulate.

Plows dig the soil to a depth of about 6–10 inches (15–25 cm). They bury everything on the surface and leave grooves, or furrows.

Cultivators dig shallower seedbeds, but are faster and less expensive to operate.

▷ A tractor pulling a plow. The cutting blades are called plowshares. They dig into the earth and turn over the soil. The earth is very dry, as can be seen from the dust.

▽ This plow, like the one on the opposite page, is a reversible type. It can be turned over at the end of the field. The plowshares now seen above ground would then turn over the soil.

◁ A rotary cultivator, or harrow, breaks up the plowed earth into smaller pieces.

▽ Machines like this disk harrow, which chops the soil up and moves it from side to side, are usually used immediately after plowing.

Plowed soil often needs further preparation by cultivators or harrows. These break the plowed earth into smaller clods. This also helps to make the surface smoother for planting or drilling.

Chemicals are often added to the soil at this stage. Fertilizers make the soil better for growing crops. Pesticides kill off insect pests that might harm the crop.

△ A machine for applying liquid manure so that it spreads evenly over the ground to act as a fertilizer. Other machines are used to apply farmyard manure, a mixture of animal waste and the straw used for bedding.

Growing the crops

Machines called drills are used for planting seeds. A grain drill cuts furrows in the soil, into which the seeds are dropped and then covered. It does this all in one operation.

For some crops, water might have to be added to the soil by irrigation machines. Weeds may be removed by harrows or cultivators, or by spraying with weedkillers, or herbicides. Pesticides are sometimes added at this stage, too.

△ A grain drill at work. The seeds are carried in the flat box, or hopper, at the back. The machine makes narrow rows for seeds, sows the seeds and then covers them over. Most drills have two hoppers so that they can apply fertilizer at the same time.

△ An irrigation machine. These huge hoses bring water to dry fields. Irrigation is used for crops such as sugar beet, potatoes and cotton, but not for cereals.

▷ A tractor can cover large areas of ground in one sweep when spraying chemicals for killing pests, weeds or diseases. Great care must be taken when spraying crops because the chemicals used can be very dangerous.

Harvesting the crops

Machines can be used to harvest almost any crop. Combine harvesters, or combines, are used for grain or seed crops such as wheat, barley and corn. They reap, or cut, the plant stalks, separate the grain or seeds and then push the remaining straw out at the back.

There are special machines for harvesting other crops, including most fruits and vegetables.

△ An "army" of combine harvesters sweeps across the cornfields at harvest time. Only very large farms have more than one or two combines.

△ The forage harvester chops the mown grass and blows it into the trailer pulled along behind it. The wheels with thin spikes deflect grass from a wider path to enable the machine to pick it up.

▷ A cotton-picker pulls the cotton from the white pods, called bolls, and collects it in the containers at the back. A machine like this can harvest as much cotton as more than 80 workers picking the crop by hand.

▷ A combine harvester unloads its grain tank into a trailer pulled by a tractor running alongside it.

The grain tank, which is just behind the driver's cab, can hold two or three or more tons of grain. The operator can see when the tank is nearly full through a glass window, or, in some machines, a light might come on in the cab. He then radios or signals the tractor to come and unload the grain. The combine harvester continues with its work while this is being done.

The arm, called an auger, through which the grain is fed into the trailer, is swung back into the combine when the operation is completed.

Different crops are harvested by all kinds of machines.

◁ Machines for harvesting blackcurrants (far left), grapes (left) and sugar cane (bottom).

▷ Harvesting a crop of green beans.

▽ A sugar beet harvester in action.

Other farm machinery

◁ Machines that pick up hay or straw and compress it into bundles are called balers. The bales, which can weigh as much as half a ton, are left on the ground to be picked up and taken for storage.

▽ Grain or seed crops are fed from the large silos to a grain dryer, the yellow structure on the right.

△ A grass seed harvester. This machine is used to encourage the grass leaves to grow. The cutting blades are set a few inches off the ground so that the seed heads are chopped off.

▷ A mower cuts the grass and leaves it ready for a forage harvester or for making hay.

△ Automatic cattle feeding stations. The cows walk into the cubicles and the feed is automatically released. In some systems, each cow has a collar round its neck with an electronic device that controls the feed and releases the correct ration.

◁ In modern milking parlors, the cows are milked by machine.

△ A hedge-trimmer in action. The hedge-trimmers that use flails, which whip around at very high speeds, are very dangerous.

▷ A hay rake is used to loosen cut grass and turn it over. This helps it to dry. Hay is dried grass. For thin crops, the rake may be used to combine two rows into one.

27

The story of farm machinery

Living off the land

Many thousands of years ago, people lived by hunting animals and gathering wild berries and seeds. They moved from place to place in search of food. Gradually, they learned how to tame animals and cultivate plants such as wheat and rice. They were able to settle down in groups and live off the land.

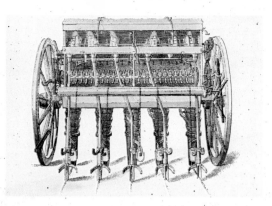

△ A drill of the 1840s, used for planting turnips and other seeds and for putting manure down at the same time.

△ An ancient Roman iron plow.

The first plows

People found that plants grew better if the earth was broken. The first plows were probably forked sticks which were dragged through the earth by hand. Crude wooden plows were developed, pulled by oxen or horses. Iron plows came into use about 3,000 years ago.

The farming revolution

New farming methods were developed in England in the early 1700s. Jethro Tull, an

English farmer, invented a horse-drawn cultivator and a seed drill. The Industrial Revolution, which began in England in the mid-1700s, brought with it a revolution in farming methods.

The mechanical revolution

A great breakthrough in the development of farm machinery came in the 1830s, with the production of a reaping machine by Cyrus McCormick,

△ A McCormick reaping machine in use in the 1860s.

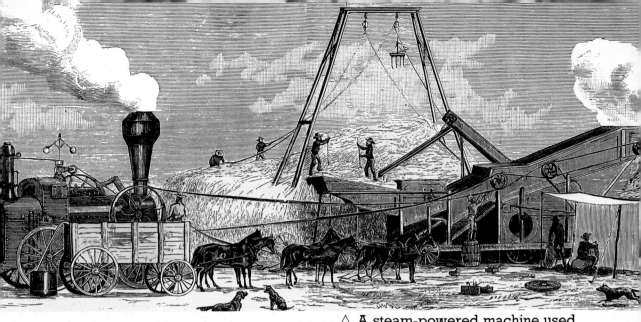

the son of a Virginia farmer. The mass production of these reapers came in time to develop the vast, rich prairie wheatlands of the United States.

At about the same time, steam-powered threshing machines came into use. A new era in farming had begun.

The tractor and the combine

Early tractors were just replacements for horses. But in the 1930s, Harry Ferguson, an Irish racing car designer, produced the TE20 and attached it to specially designed implements, such as plows. By the 1950s, the TE20 was being used all over the world.

The other great modern invention was the combine harvester. This was developed over a long period, especially by Massey-Harris in Canada. It came into widespread use in the 1950s.

Farming by computer

Big advances are being made by the use of computers. Many modern combines and tractors have them to help the operator. They are also used in livestock farming, for example, to control milking and feeding.

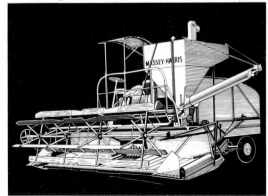

△ The MH 20, built in 1938, was the first successful combine.

29

Facts and records

△ The world's biggest baler at work.

Biggest bales

Bales are round or rectangular and come in all sizes. Small bales weigh from 55 to 110 lb (25 to 50 kg). Round bales might weigh up to half a ton. The largest bales of all are the rectangular ones produced by the Hesston 4800 baler. These measure 4 × 4.2 ft square and up to 8 ft long (1.22 × 1.29 m × 2.44 m). A hay bale of this size would weigh about a ton.

A curious machine

One of the most curious-looking farm machines was the Massey-Harris stripper-harvester built in about 1900. It had the appearance of a bicycle and sidecar. It was towed, but had its own motor to operate the cutting and threshing mechanisms. The combine harvester was developed from machines like this.

△ The Massey-Harris machine.

Glossary

Baler
Machine for compressing hay or straw into regular bundles.

Combine harvester
Machine that reaps and threshes grain and disposes of the stalks in one operation.

Crawler
Tractor with steel tracks instead of wheels.

Cultivator
Machine such as a plow or harrow for digging up and turning over the soil.

Drill
Machine for planting seeds.

Fertilizer
A special substance or material added to the soil to help plants grow.

Forage
A crop used for cattle feed.

Harrow
Machine for breaking up plowed soil into smaller pieces, or clods.

Hay
Dried grass.

Herbicide
Substance for killing weeds.

Manure
Mixture of straw bedding and animal waste used for fertilizing the soil.

Pesticide
Substance for killing pests that destroy crops.

Plow
Machine for digging up and turning over the soil.

Reaper
Machine that harvests a crop.

Silage
Wet grass that is stored for later use as animal feed.

Silo
Storage tower.

Straw
Stalks of grain after threshing.

Thresher
Machine that separates the grain from the ear.

Tractor
Vehicle that pulls or operates other machines or implements.

Index